WITHDRAWN

CHEROKEE

JAN 6 - 1992

Our World

POLLUTION AND CONSERVATION

Malcolm Penny

Silver Burdett Press
Englewood Cliffs, New Jersey

Titles in this series

Coasts

Deserts

Grasslands

Jungles and Rainforests

Mountains

Polar Regions

Pollution and Conservation

Rivers and Lakes

Seas and Oceans

Temperate Forests

First published in 1988 by
Wayland (Publishers) Ltd
61 Western Road, Hove
East Sussex BN3 1JD, England

Adapted and first published in the
United States in 1988 by Silver Burdett Press,
Englewood Cliffs, New Jersey

© Copyright 1988 Wayland (Publishers) Ltd

© 1988 this adaptation Silver Burdett Press

Edited by Jollands Editions

U.S. edition edited by Nancy Furstinger

Series design by Malcolm Smythe
Book design by Tony Truscott

Typeset by DP Press, Sevenoaks, Kent
Printed in Italy by G. Canale & C.S.p.A., Turin

Library of Congress Cataloging-in-Publication Data
Penny, Malcolm.
  Pollution and conservation
  (Our World)
  Bibliography: p.
  Includes index.
  Summary: Describes the various kinds of pollution and what is being done to protect the natural environment.
  1. Pollution—Environmental aspects—Juvenile literature.
2. Nature conservation—Juvenile literature. [1. Pollution.
2. Man— Influence on nature. 3. Conservation of natural resources.
4. Environmental protection] I. Title. II. Series.
QH545.A1P46 1988 363.7'3 88–31856
ISBN 0–382–09792–0

Front cover, main picture Pollution from a power station in Slough, England.

Front cover, inset A Southern right whale, so called because it was once considered the "right" whale to hunt. Right whales are now protected in most countries.

Back cover Rainforest clearance in Malaysia.

Contents

Chapter 1　Pollution and people	
The origins of pollution	4
Land pollution	6
Noise pollution	8
Chapter 2　Air pollution	
Smog and car exhausts	10
Acid rain	12
Nuclear problems	14
Pollution and the ozone layer	16
Chapter 3　Pollution in water	
Freshwater pollution	18
Pollution in the sea	20
Chapter 4　Conservation	
Our changing planet	22
The conservation of rainforests	24
Estuaries and wetlands	26
The conservation of grasslands	28
Conservation on islands	30
Chapter 5　International conservation	
Migrants	32
Cooperation across frontiers	34
National and international parks	36
Chapter 6　Conservation and people	
Waste not, want not	38
Populations	40
Antarctica: A clean continent	42
Glossary	45
Further reading	46
Index	47

CHAPTER 1: POLLUTION AND PEOPLE

The origins of pollution

Pollution occurs when harmful substances are released into the environment, causing damage to living things. Any discarded material which cannot be broken down by natural means, for example plastic bottles or old machinery, also can be considered a form of pollution.

Pollution is nothing new. The people of the past spoiled their surroundings in the same way as we do today. However, two things have changed. These are the nature of the pollution and people's attitudes toward it.

In cities a century ago, the common forms of pollution included horse-manure in the streets, poor sanitation, and thick billows of smoke spreading across the landscape.

Today, the nature of pollution has changed. Horses have disappeared from the streets, and many cities are smokeless zones. Sanitation has

improved, but there are still serious threats to world health. The new threat of radiation has joined other modern pollutants such as lead from gasoline, oil dumped or spilled into the sea, and poisonous chemicals used in agriculture.

Litter was less of a problem when the materials that were thrown away were biodegradable, that is, they rotted naturally and were absorbed into the ground. Modern plastic wrappers and containers do not rot away, but spoil the landscape for many years unless they are disposed of properly. If they are burned, the fumes that they produce may themselves become another source of pollution.

As people multiply, so does the pollution that they create. In Kenya, the fastest-growing country in the world, the population is expected to double in less than twenty years. In that time, the country will have to produce twice as much food, double its industry, and dispose of twice as much sewage and other waste. The same problems, on a lesser scale, face many other countries of the world.

In this book, we shall look at ways of reducing the main forms of pollution and how to lessen their effects. We shall look also at ways of reducing the effect of an exploding population on the finite world in which we and our descendants must live.

Above Even after World War I cities were polluted by smoke from factories. This is a cotton mill in Preston, England, in 1925.
Below Today, people such as these marching in a demonstration are more aware of the dangers of pollution.

Left Almost every human activity causes some form of pollution: 1 Burning fossil fuels to produce energy leads to the formation of 2 acid rain. 3 Dumping garbage can cause the pollution of underground water supplies. 4 Chemicals that run off farmland find their way into the water table. 5 Poorly treated sewage pollutes rivers. 6 Oil spillages from ships pollute the seas and oceans.

POLLUTION AND PEOPLE

Land pollution

As the world population increases, so does the demand for food. This places a great strain on farmers to produce more from the land. Overworked farmland can be destroyed by soil erosion, careless irrigation, and poor plowing. Each year millions of acres of farmland in the world are lost.

Some of the steps which farmers have taken to grow more crops have led to pollution problems.

As crops grow, they remove important chemical nutrients from the soil. The nutrients, which are the crops' food, must be replaced to keep the soil fertile, or rich. In traditional farming, the nutrients are replaced by animal manure, by "resting" the land between crops, and by rotating the crops, which means growing a different crop in a field each year. Modern farming is based on adding artificial fertilizers to the soil. These chemical fertilizers make it possible to grow very large crops and to use the land every year. This can cause problems when the chemicals run off the land and upset the balance of life in watercourses.

To protect their crops, modern farmers spray them with weed killers and pesticides. At first, these chemicals were regarded as major breakthroughs. Then people realized the harm that some of them were doing to the environment. DDT and its relation dieldrin were the first to be proved harmful. These insecticides stayed in the soil, killing insects other than the pests. Eventually, they worked their way up the food chain, killing birds of prey, otters, and other harmless creatures. The chemicals also had serious toxic effects upon some of the people who worked with them.

The use of such pesticides has now been banned in the developed industrial countries. Sadly, they are still being used in parts of the developing world.

Every day, modern cities produce many tons of garbage.

Above The concentration of pesticides increases in creatures higher up the food chain. When a mouse eats insects which have been killed to protect corn, the poison collects in its body. A hawk that eats several poisoned mice may take in a dose which could kill it.

Pesticides and fertilizers increase crop yields, but more land is always needed to feed the growing population.

POLLUTION AND PEOPLE

Noise pollution

The screech, thump, and rattle of a factory floor in full production was once looked upon as a sign of progress. Even in this century, the workers in mills and steel foundries had to learn to lip-read to speak to each other.

Noise pollution is a relatively new idea, but it is taken very seriously. Industrial employers now have to provide ear protectors for their workers wherever the noise of machinery might be harmful. In the street, cars, motorcycles, and trucks must be fitted with mufflers. This keeps the noise they make down to acceptable levels. Even so, to stand close to a heavy truck can be very unpleasant.

Pop concerts are exciting because they are so loud. It is now known that noise at such a level can cause permanent damage to the hearing of the audience, as well as to the members of the band.

Some noises are not physically harmful, but annoying to those who are not involved. In the country, some agricultural processes, such as drying grain or combine harvesting, disturb the peace for short periods. The "noise lobby" is very active in its efforts to have the noise levels reduced, even in important activities like farming.

Air travel is another necessary and noisy part of modern life. People living near airports have to put up with the constant noise of airplanes. Modern jet planes are much quieter than earlier designs. Their flight paths and movements in and out of airports are controlled carefully. This helps to reduce the annoyance that they cause to people on the ground.

In the same way, the use of portable radios in public places, which to many people is the audible equivalent of litter, is controlled by regulations.

Pop concerts can produce higher noise levels than the engines of a jumbo jet at full throttle.

Right Airplanes are restricted in their movements at night, but during the day they can make life very unpleasant for people who have to live and work near airports.

Below right People can choose not to expose themselves to the harm that very loud music can cause; but few can escape the constant noise of traffic in a city.

Description of noise	Noise level (decibels)
Silence	0
Whisper	10
Medium-sized office	55
Conversation (3 ft. apart)	60
Traffic on busy street	70
Motorcycle accelerating	85
Discotheque	95
Boeing 747 taking off	108
Pop concert	110

CHAPTER 2: AIR POLLUTION

Smog and car exhausts

The word "smog" was invented to describe the mixture of smoke and fog which once blanketed great industrial cities like London. In December 1952, a four-day smog in London killed more than 4,000 people. It remains one of the greatest air pollution disasters the world has ever seen. Since then, laws have been passed which allow only smokeless fuels to be burned in cities.

Today, smog has a slightly different meaning. In cities where smog is a problem it is partly caused by fine weather. Strong sunlight reacts with vehicle exhausts and industrial fumes. This creates a photochemical haze, which can be very harmful to people's eyes and lungs. In Los Angeles and Tokyo, for example, people are advised not to go outdoors on days when the smog is particularly bad.

Some of the worst air pollution in the world is in Mexico City. Passengers in approaching airplanes can see the smog hanging over the city. As the air enters the cabin before landing, the passengers can actually smell the Mexico City smog.

A common form of air pollution in city streets is carbon monoxide. Carbon monoxide is produced by car exhausts and is highly poisonous. It causes changes in the blood, reducing its ability to carry oxygen throughout the body. In small quantities, its effect is merely to make people feel tired and short of breath, but in larger quantities, it kills.

Lead is another serious pollutant from car exhausts. It is added to gasoline to make engines run more smoothly. Lead is known to be dangerous, mainly to growing children. Their nervous systems and intelligence can be seriously damaged by it.

Lead-free gasoline has been available to motorists for years, and many new cars are designed to run only on unleaded fuel. In several states the use of leaded gasoline is illegal. Despite its availability in Great Britain, only one in a thousand motorists used lead-free gasoline in 1987.

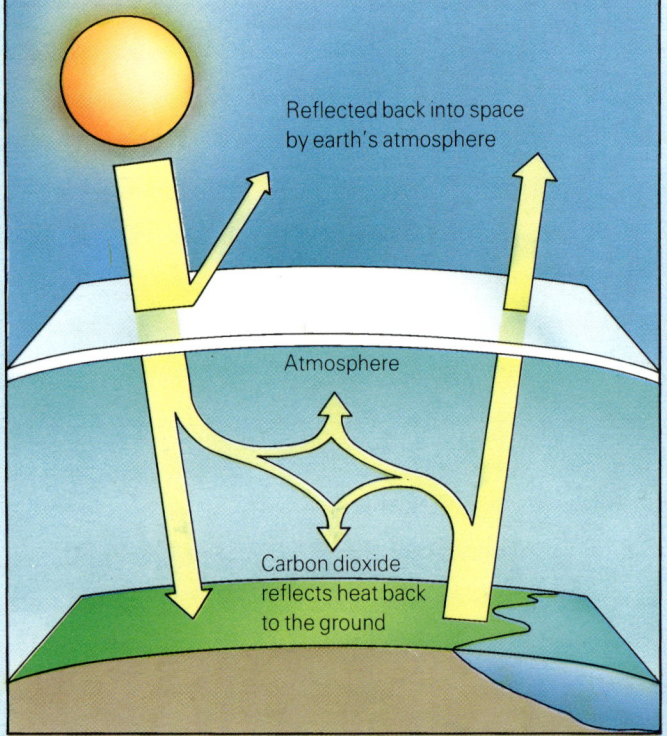

The Greenhouse Effect
The atmosphere absorbs or reflects half the energy that reaches the earth from the sun. Carbon dioxide tends to trap or reflect back to the ground heat that would otherwise escape back into space. Burning fossil fuels and clearing forests by fire have increased the carbon dioxide content of the atmosphere: it has risen by 30 percent since 1850. As a result, the earth is becoming warmer. Scientists have calculated that by the year 2060 the amount of carbon dioxide will be twice what it was before 1900. This could raise surface temperatures by 34°F at the equator, 37°F in temperate latitudes, and by as much as 45°F at the poles. The pattern of rainfall would change, and sea levels would rise. Areas that now grow the most grain would produce less, and places that are at present almost deserts could become fertile. The map (**opposite**) shows the principal grain-growing areas, some of which would become too dry to continue, and the places that would become wet enough to grow food.

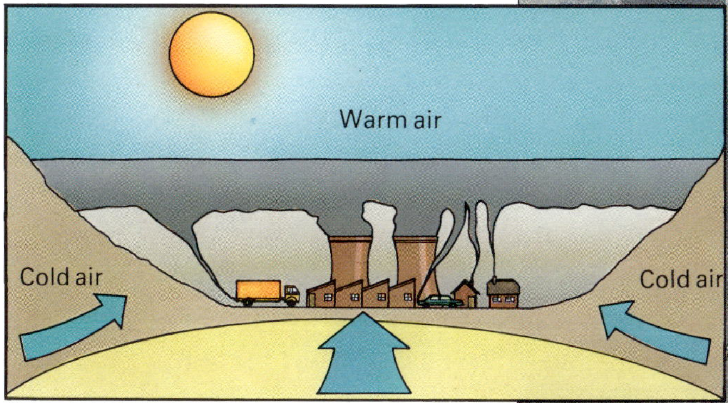
Warm air
Cold air
Cold air

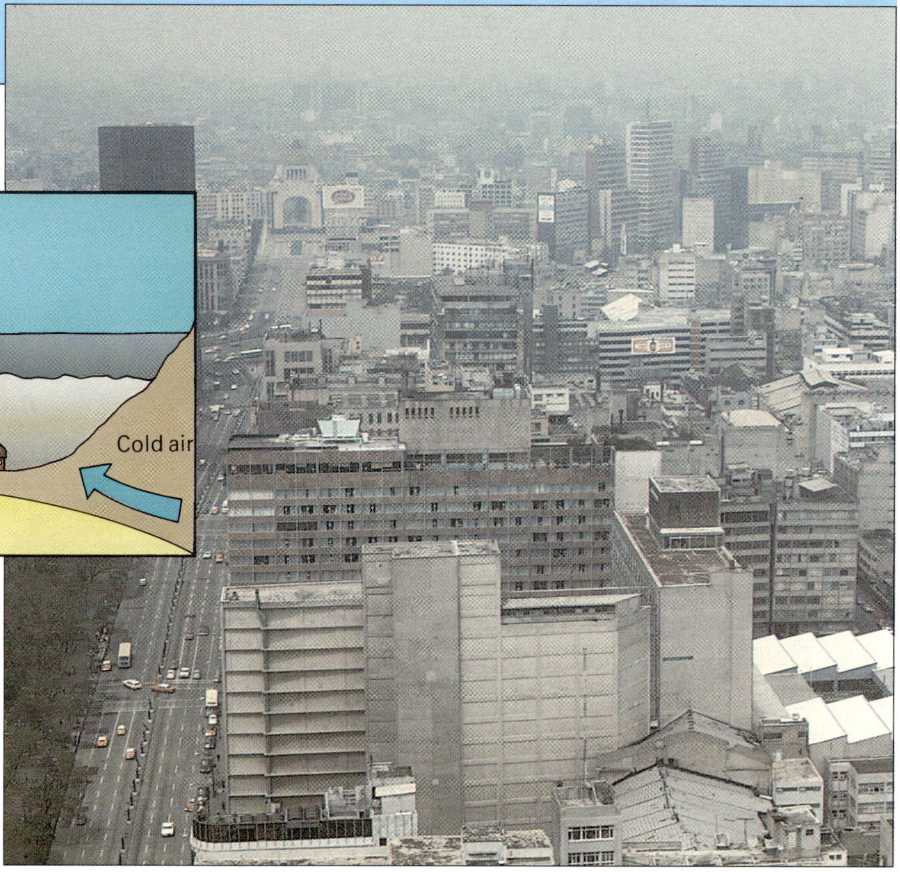

Above Smog forms in a city when warm air traps cool air beneath it. The cool air becomes polluted with exhaust fumes, which react with sunlight to form harmful chemical compounds.

Right Smog in Mexico City.

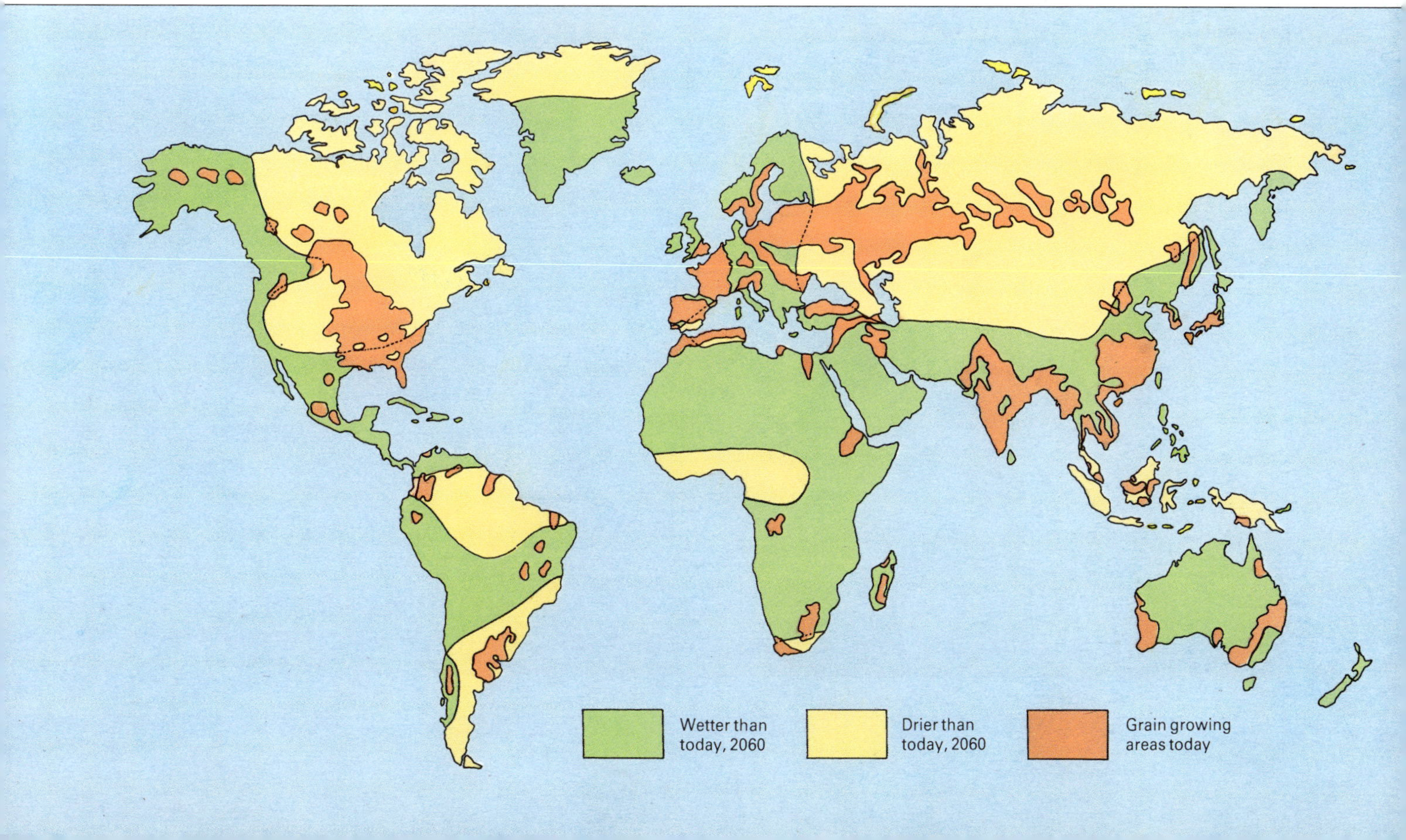
Wetter than today, 2060
Drier than today, 2060
Grain growing areas today

AIR POLLUTION

Acid rain

All rain contains some weak acids. This is because it reacts with the gases which occur naturally in the air, such as carbon dioxide and sulphur dioxide. Since about 1960, however, people have noticed that rain has become more and more acid, to the point where it is damaging buildings, crops and trees, and aquatic wildlife.

The "new" acid rain has been blamed on industrial pollution. The precise way in which air pollution kills trees is still being studied by scientists. Most scientists agree that the problem is man-made.

When fossil fuels are burned, they produce sulphur and nitrogen, which react with damp air to make sulphuric and nitric acids. If the air is dry, the reaction takes place very slowly. The rain might not fall until the fumes have drifted into damp air far away from the place where they were produced.

This has created an international problem, since fumes from factory chimneys in the United States, for example, blow northward and may fall as acid rain over Canada. A similar problem exists in Great Britain, where fumes from factories are blown across Europe or Scandinavia. There, lakes and rivers have become acid, damaging plant life and killing fish. Large areas of soil also have been affected, so that forests are dying.

The damage has been very serious. For example, in 1900, anglers in Norway caught more than 66,000 pounds of salmon, but since 1970, they have caught none. In Canada, where the trout and salmon fisheries are a major industry, many of the lakes and rivers have no fish left in them.

The famous Canadian maple trees are dying as well. Acid rain may damage forests by removing vital elements from the soil, so that the trees cannot take these nutrients up in their roots. The acid may also affect the leaves of some trees, reducing their ability to make food by photosynthesis. As a result, the trees starve. At the same time, they become more at risk from severe frosts and pests.

The latest research into the death of trees suggests that there is no single cause. Scientists believe that the trees may be dying from a mixture of tens or even hundreds of chemicals which are released into the air.

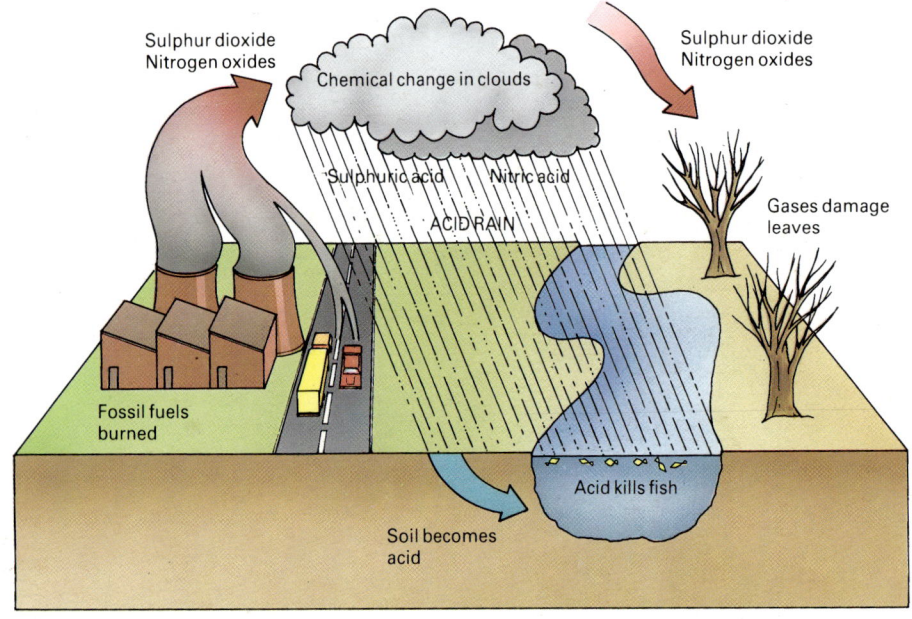

Left Burning fossil fuels — in homes and vehicles as well as in factories and power stations — produces fumes that contain sulphur dioxide and oxides of nitrogen. These substances react with damp air to produce sulphuric and nitric acids. When the rain eventually falls, it affects all forms of life.

Right The damage to trees can be seen in large wooded areas such as the Black Forest in Germany.

AIR POLLUTION

Nuclear problems

Nuclear power stations do not produce the pollution caused by burning fossil fuels, but they present other problems. The radioactive fuel that they use is very dangerous if it escapes into the air, water, or soil. The radiation it gives off, even in small amounts, can damage human cells, causing cancers in anyone who is exposed to them. It also causes deformities in unborn babies. Large amounts of radiation are deadly.

Although every precaution is taken, accidents are bound to happen. There have been two very serious disasters in nuclear power stations. One was in 1979, at Three Mile Island in Pennsylvania, and the other was in 1986 at Chernobyl in the Soviet Union. Smaller accidents in Great Britain and other countries have shown that nuclear power stations cannot be made completely safe.

Another problem is to find a way of disposing of nuclear waste, including used fuel, which is still radioactive. Various methods have been tried. These include burying the waste in concrete containers deep beneath the sea, or pumping it into disused mine shafts. Whatever methods are used, the material will continue to be dangerous for many years to come. Some forms of waste will be radioactive for centuries.

Below right In spite of precautions, nuclear accidents are inevitable. At Chernobyl, an explosion wrecked the main reactor (arrow).

Below In this experimental train crash, the yellow nuclear waste container was undamaged.

Waste from a nuclear power station can be stored underground, cast into blocks of solid glass. After 70 years, the blocks can be sunk in the sea, but they will not be safe for a thousand years.

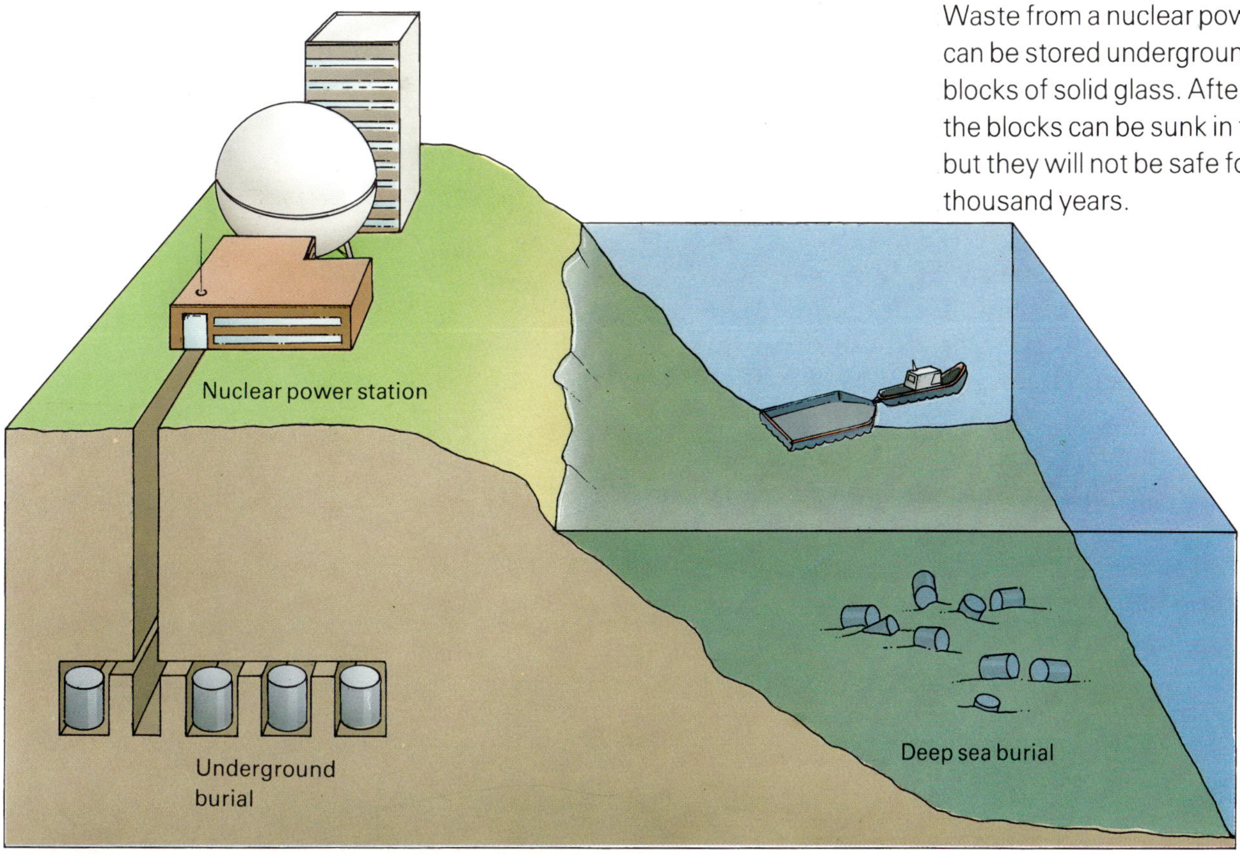

Nuclear power station

Underground burial

Deep sea burial

The danger from nuclear fuels and wastes has to be balanced against the advantages of nuclear power. The world needs alternatives to fossil fuels, which will run out in the future. Those in favor of nuclear power say that it is relatively safe. Many more people are killed every year in coal mines and on oil rigs than in nuclear accidents.

Pollution from radioactive material is not all man-made. There is a certain amount of low-level, natural radiation all around us. This is in the form of radon gas which is released from granite rocks all over the world. This "background radiation" is thought to cause minor damage to people, but it is not a serious problem. Those people who are opposed to nuclear power argue that humans have no right to add to that damage, harming not only themselves but future generations.

AIR POLLUTION

Pollution and the ozone layer

The ozone layer is a region in the earth's stratosphere — between 7 and 30 miles above the surface. Ozone is a form of oxygen which is created by the action of sunlight. It absorbs much of the harmful ultraviolet (UV) radiation from the sun. Ultraviolet radiation is one of the causes of skin cancer.

Since 1979, scientists have noticed that the ozone layer is becoming thinner in places. The thinning is believed to be caused by a group of chemicals called chlorofluorocarbons (CFCs), which are used in aerosol spray cans, air conditioners, refrigerators, and other common articles, including the polystyrene food containers used for fast foods. When they are released, CFCs rise to the stratosphere, where sunlight converts them into chlorine monoxide. The chlorine monoxide reacts with ozone, removing it from the air.

Below Sunbathers, like these in Australia, are protected from the worst effects of UV radiation by the ozone layer.

Left Visible light passes straight through, but most invisible UV radiation is absorbed by the ozone layer. If this layer becomes thinner, more UV will reach the surface.

Right Many common articles contain CFCs, which are damaging the ozone layer. These include coolants in refrigerators and air conditioners, propellants for aerosols, and foaming agents for heat-proof polystyrene food and drink containers.

The seriousness of this effect was first realized in the autumn of 1985, when a hole appeared in the ozone layer over Antarctica. The hole has been seen every summer since then, and some scientists fear that it will spread to lower latitudes, nearer to the equator.

If more UV radiation reaches the earth, the increase of skin cancer may not be the only result. Grain crops could become less productive and the UV could warm the atmosphere, adding to the "greenhouse effect" caused by rising amounts of other pollutants such as carbon dioxide and methane.

In the interest of preserving the ozone layer, the United States banned the use of CFCs in spray cans. Berkeley, California, was the first city to ban fast food containers made of polystyrene, in 1987. Other cities will certainly follow. However, even if CFCs were banned immediately all over the world, the effects on the ozone layer will continue for some time. As old refrigerators and air conditioners are scrapped, they will release the damaging gases into the stratosphere.

CHAPTER 3: POLLUTION IN WATER

Freshwater pollution

The pollution of fresh water affects all forms of life, from the smallest plants to large animals, including humans.

Water is polluted mainly by sewage, industrial waste, mines, agricultural chemicals, and agricultural waste. Even farmyard manure washed off the land into streams and ponds can cause the effect known as "eutrophication," or "over-feeding." This results in a massive growth of bacteria and small plants. As they multiply, they use up all the oxygen in the water. In time, all the water creatures die from lack of oxygen.

Artificial fertilizers create the same effect. Nitrates, which are artificial fertilizers, are used increasingly in all agricultural countries. Trickling through the soil, they seep through rocks and into the ground water. This water supplies wells, so eventually nitrates find their way into tap water. It is a worrying thought that nitrates can cause stomach cancer in some animals. Their effect on humans is not known for certain.

Pesticides and weed killers also cause water pollution. This happens if they are washed into streams after being used on land. The pesticide methyl mercury is used to protect seeds against bacteria and insect pests. However, it kills not only birds which eat the seeds, but also fish and other water creatures in streams and ditches. Dieldrin is an insecticide that has been banned for most purposes. It is still used for treating wood and mothproofing blankets, and is occasionally found in polluted ground water.

New chemicals invented for industry create new problems as they escape into the environment. Among the latest are strong cleaning fluids called polychlorinated biphenyls (PCBs). Trichloroethylene, which is used to clean engines, and tetrachloroethylene, used in tanning leather, are two PCBs that find their way into the ground water when they are spilled or thrown away carelessly. Their immediate effect is on freshwater life. In time they, too, are found in tap water.

Left Agricultural waste from the local farm has upset the balance of life in this pond in Oxfordshire, England. The waste, rich in plant foods, has caused widespread growth of small water plants called algae.

Right A spill of tannic acid, used in dyes for tanning leather and making ink, has severely polluted this river.

POLLUTION IN WATER

Pollution in the sea

People have always used the sea as a dumping place. When the world population was smaller, and the rubbish was biodegradable, the effect was unnoticeable. Many people today act as if that were still the case.

However, people are beginning to realize that the sea has its limits as a universal dump. Even large oceans, such as the Atlantic and the Pacific, are badly contaminated. Many beaches along the Atlantic Coast were closed in the summer of 1988 because of illegally dumped hospital wastes. Oil, PCBs, and dieldrin have been found in the Antarctic Ocean. The smaller, shallow seas, such as the Mediterranean and the North Sea, are in danger of becoming poisoned beyond recovery. All oceans are threatened by plastic debris, such as discarded fishing nets that endanger marine life.

Although there are some discharges of oil and other pollutants directly into the sea, most marine pollution is first discharged into rivers, from farmland, factories, and towns. In Europe, fish in the North Sea are being affected by pollution. The pollution comes from agricultural and industrial areas far inland in Germany and the Netherlands. Great Britain and other countries whose coasts are washed by the North Sea also add to the problem.

Other activities far inland can affect marine life. Forest clearance and mining can both cause erosion. Large quantities of fine silt are washed into rivers and carried to the sea. The silt can smother coral reefs, and other life on the seabed.

To protect the hulls of boats against barnacles and algae, they are coated with paint containing a pesticide called tributyl tin (TBT). This gradually

The main sources of pollution in the North Sea.

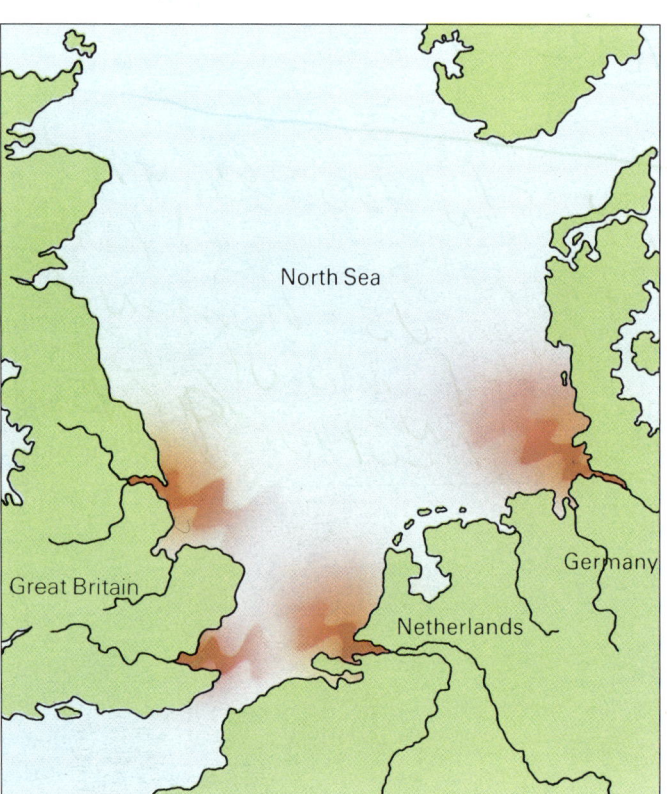

Volunteers clean birds caught in an oil spill.

20

Above Oil pours from a sinking tanker, the *Amoco Cadiz*.

Below Untreated sewage is steadily polluting the sea.

seeps into the sea, where it contaminates shellfish, and even salmon. The risk of harm to humans has finally alerted people to the dangers of using TBT.

A rare form of pollution is caused by the warm water which flows from the cooling systems of power stations. This "thermal pollution" can change the way of life of marine creatures. For example, in Florida manatees have begun to gather around a power station outflow rather than migrating to warmer waters in winter. If there is any failure in the output of warm water, the manatees die of cold.

CHAPTER 4: CONSERVATION

Our changing planet

The growth of the human population is changing the world in which we live. In addition to causing the pollution of the environment, human activities have led to the reduction in numbers of many species of wildlife. This can be blamed partly on people hunting animals for food and sport. It also is due to the spread of agriculture and human settlement. This has severely limited the areas where wild animals can live.

When people traveled to new lands, they took with them plants and animals from their homeland. This happened when Europeans first settled in Australia, 200 years ago. Rats, mice, deer, and rabbits were all introduced—accidentally or otherwise—and their numbers spread rapidly. They soon overran the country, destroying vast areas of pasture. This had disastrous effects on the native wildlife, which had to compete for food.

All forms of wild habitat have been affected by the pressure of people, who need space in which to build cities and grow crops. By cultivating prairies, draining wetlands, and felling forests, people have made irreversible changes to their environment. Some of these changes have led to human disasters, such as the famines in parts of India and northern Africa, and the creation of the dust bowls in the United States.

Overcultivation, without proper irrigation or the use of enough fertilizer, can create deserts out of what was once productive farmland. Where rainforests are felled to provide land for growing crops, the result is similar. Without the trees to provide leaf litter and to hold the soil in place, the land soon becomes eroded and lifeless.

Human activity is responsible for other "natural" disasters. The floods that regularly overwhelm the low-lying areas of Bangladesh have two direct causes. One is the felling of mountain forests far inland among the foothills of the Himalayas. Soil washed from the hillsides silts up watercourses, which causes flooding on the plains below. The other reason is the clearing of the mangrove forests which once protected the mouths of the rivers from the sea.

The human activities causing such problems need to be better understood and controlled. Conservation of the natural world is the most urgent problem facing humanity today.

Left The tragic drought and famine in Ethiopia was partly caused by human activities. Huge herds of cattle removed all the ground vegetation by overgrazing, and an ever-growing human population burned the trees for fuel. When the rains failed, the country became a desert.

Topsoil washed downhill

Forest clearance

Flood plain

River bed silts up

Above Rain falling in the mountains sweeps down the river without forested foothills to store it and release it gradually. It brings along topsoil which blocks the lower part of the river, causing floods on the plains.

Right Flooding in the Ganges Delta in Bangladesh.

CONSERVATION

The conservation of rainforests

The tropical rainforests are richer in plant and animal species than any other habitat on earth. This suggests that they are very fertile places, ideal for growing food, but they are not.

The soil is much poorer in quality than in other forests. The richness of the rainforests, and even the rainfall, depends on the trees. This is because most nutrients are absorbed by the roots of the huge trees. Also, the trees act like sponges, soaking up water and passing it out through their leaves. Once the trees are gone, the amount of rainfall decreases, and the soil soon loses its nutrients.

Small groups of people are able to grow food in rainforests. For example, in Central America the traditional method of "slash and burn" agriculture is used. First, an area of forest is cleared. Then crops are grown for a few years, until the soil loses its fertility. The ground is abandoned and left to recover, for ten or twenty years. However, in most countries there are too many people to leave the land rest long enough, and as a result the forests are destroyed.

When rainforests were first commercially exploited, it was for their valuable hardwood timber, such as teak and mahogany. The timber was "clear-cut," leaving nothing behind. Vast areas were completely destroyed, leaving the wildlife homeless.

People in Sydney hold a protest against the destruction of Australia's rainforests in the Daintree region of North Queensland.

24

Above and below Amazonian Indians know medicinal uses for many of the plants in the forest. When the forest is cleared, these herbal medicines will be lost.

Some people might say that the loss of species such as orangutans and sloths is not important compared with the needs of people. However, the rainforests contain a huge variety of unknown plants. Among these, new crops and medicines certainly remain to be discovered.

The conservation of the rainforests is based on the idea of "sustainable use." This means exploiting the forests in a controlled way, so that they are able to recover. Trees for timber can be selectively felled, by choosing the most suitable ones and leaving the rest to grow.

Small clearings scattered throughout the forest are much better than large ones. This is because they do not interfere with the rainfall or the fertility of the soil in the same way that large clearings do.

CONSERVATION

Estuaries and wetlands

An estuary, where a large river flows into the sea, is a very challenging place for wildlife to live. The water regularly changes from fresh to salt, and large areas of mud flats are exposed at low tide. The few animals and plants that are adapted to living there do so in enormous numbers, because they have so little competition. Estuaries are very important sources of food, especially for wading birds, ducks, and geese.

Estuaries are threatened by several human activities as well as pollution. The most serious is land reclamation, in which parts of the estuary are drained and enclosed by banks. The reclaimed land can be used for agriculture or housing. If a sea wall or a harbor is built at the mouth of an estuary, the changed action of the waves and tides may alter the estuary greatly.

Estuaries also are damaged by changes in the amount of water flowing down a river. This may happen because some of the water has been taken for human use or irrigation, or because the river has been deepened to make it navigable.

The threat from a tidal barrage is even greater. A barrage is a dam built across a river mouth, so that the rise and fall of the tide can be used to generate electricity. A barrage may also be built to control flooding. By altering the flow of salt water into an estuary, and so allowing fresh water to build up, a tidal barrage may change the nature of the estuary's ecology completely.

"Wetlands" are other watery habitats. They include marshes, bogs, and swamps, all of which are important to wildlife, especially migrating birds. Like estuaries, freshwater marshes are threatened by drainage and pollution. The drainage need not be aimed at the marsh itself. Often, when nearby rivers are dredged, the water table falls so low that the marsh dries out. Conservation of marshes sometimes involves pumping water into them during periods of dry weather.

The conservation of all wetlands, like that of rainforests, involves a careful balance between the needs of humanity and the survival of wildlife and the natural environment.

Protected wetlands, like this estuary in New South Wales, Australia, provide a haven for migratory birds.

Two ways of destroying wetlands. Thoughtless dredging and clearing vegetation from the banks (**left**) can turn rivers into lifeless drainage canals. Here, land in the Everglades, in Florida, is being drained for building. Plowing marshlands (**right**) in Sutherland, Scotland, cuts drains which provide drier land on which trees can be planted. At the same time, it destroys one of Europe's most ancient peat bogs — rich in rare plants and animals, and an important nesting place for wetland birds.

CONSERVATION

The conservation of grasslands

The great grasslands of the world—such as the Serengeti plain in East Africa, the pampas of Argentina, or the steppes of Russia — may seem to be very stable and secure habitats. However, what could become of them is exemplified by the prairies of North America, which are now covered in cities and agricultural land.

Mighty herds of buffalo once roamed across a sea of grass, but they were slaughtered by the thousands by hunters in the last century. When the land was clear, the settlers moved in, set up their ranches, and built their farming towns.

The American prairies must have seemed endless to the first explorers, as the steppes and the African plains still seem today. But now people are aware that all the earth's resources are finite, and must be conserved for the future.

Uncontrolled grazing by cattle is a serious threat to grasslands. Overgrazing results in the ground becoming bare. Without plant cover, the soil is easily eroded and the land becomes a desert. In the dry lands to the south of the Sahara Desert, overgrazing causes human disasters on a vast scale. Even in countries with reliable rainfall, overgrazing may destroy rare plant species and cause erosion.

In Australia, irrigation improved the grazing for the early settlers. It also upset the ecology of the outback by increasing the populations of the large kangaroos. The kangaroos became a serious nuisance to the farmers.

The new grazing land had to be used with care. The density of cattle in the Australian outback is measured in acres per animal, rather than the other way around. By controlling the grazing, the farmers conserve the grasslands.

The value of grasslands lies not only in their use for cattle and sheep. They also provide a valuable source of plant species from which new crops might one day be developed. All the cereal crops that we grow are descended from wild grasses.

With carefully controlled grazing, even the dry grasslands of South Australia can support cattle.

Right Without weed killers, a field contains many different plants.

Below Wheatfields contain only one species, a form of grass.

Below Overgrazing of the American prairies led to the creation of a dust bowl in the 1930s.

CONSERVATION

Conservation on islands

Most animals and plants that live on remote islands have evolved to be different from those anywhere else in the world. Because of this they are of great interest to naturalists. In the 1850s, Charles Darwin developed his theory of evolution after he had studied the animals of the Galapagos Islands, off Ecuador, in South America.

The balance of life on an island is easily upset. The greatest threats to island wildlife come from the plants and animals that humans introduce. For example, when Europeans settled in New Zealand, they introduced the common bramble bush, which soon spread out of control. Animals such as goats are especially dangerous, because they can destroy the native vegetation. Other animals, like cats and rats, may prey on ground-nesting birds.

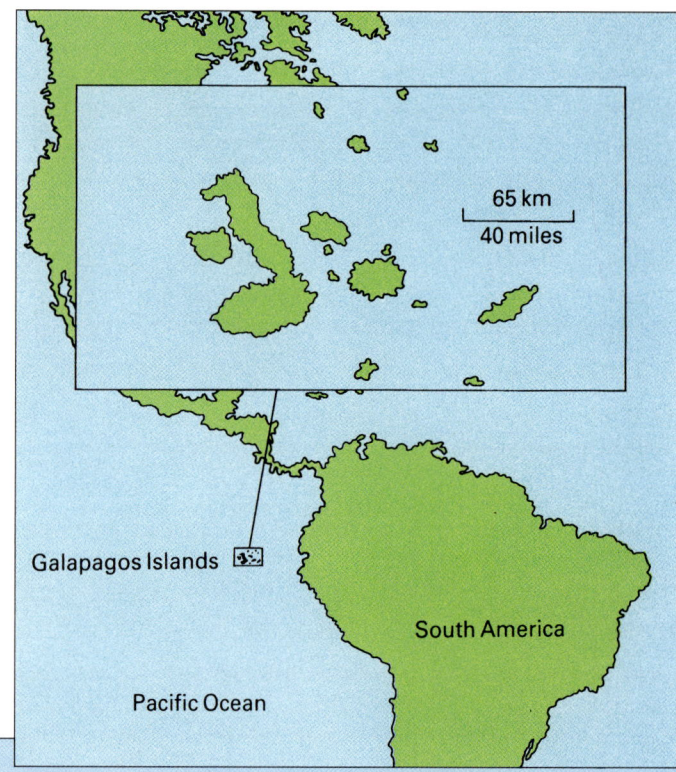

Giant tortoises can survive only where they do not have to compete with mammals, as in the Galapagos Islands.

Introduced predators are able to succeed so well because island animals usually have little or no defense against them. Many flightless birds were wiped out when people and their animals settled on their island homes. The dodo and the solitaire, which lived on islands in the Indian Ocean, are famous examples. In New Zealand, the moa became extinct after the Maoris arrived. The other flightless birds such as the kiwi, the kea, and the kakapo were safe until Europeans introduced cats and dogs. There are still a few of these species left, and they are being carefully protected.

The conservation of the few remaining unspoiled islands must take the form of excluding all new and exotic animals and plants. On most other islands, where the introductions have already happened, people now try to remove or at least control the spread of non-native species.

The study of islands and their conservation is very important. Elsewhere in the world, game reserves and national parks are becoming increasingly like islands, cut off from each other in a sea of cultivated and populated land.

The Seychelles fody (**above**) evolved in the islands in the Indian Ocean. It now lives alongside the Madagascar fody (**below**), which was introduced by early settlers.

Right The kiwi is typical of many of the birds of New Zealand. It evolved in a place where there were no mammals to prey on it or to compete with it for food. As a result, it lost the ability to fly, and evolved into a nocturnal bird of the forest floor, feeding on earthworms and beetles, as if it were a mole or a mouse. Unlike other birds, its nostrils are at the tip of its bill.

CHAPTER 5: INTERNATIONAL CONSERVATION

Migrants

Migrating birds regularly cross international boundaries. Their survival depends on the protection of feeding and resting places by each country along their route. For example, swallows which breed in the northern temperate zone spend the winter far to the south. As they move between their breeding and wintering grounds, they become the responsibility of a series of different countries. Their conservation demands international co-operation, which is not always easy to arrange.

Below Sandhill cranes migrate from Canada and the western Rockies to spend the winter in New Mexico.

Overpopulation in Mexico has produced a threat to migrant birds as well, as people move from the lowlands to try to raise crops in the mountain forests. Because forest land does not remain fertile for long after being cleared, more and more trees are felled every year. Many North American songbirds spend the winter in the Mexican forests. Although their summer breeding places are carefully protected, the birds are dying in Mexico because their winter homes are being destroyed.

The only solution to this problem will be for the hungry people to find a way to grow crops in the forests without destroying these areas.

Left Birds in Thailand are often trapped to be sold as pets in city markets. Among them are migrating birds from other countries. These birds in Bangkok will never migrate again.

Above Forest cleared for farming no longer offers shelter to wintering monarch butterflies.

Below Most monarch butterflies migrate to a small area of Mexico. A few winter in California.

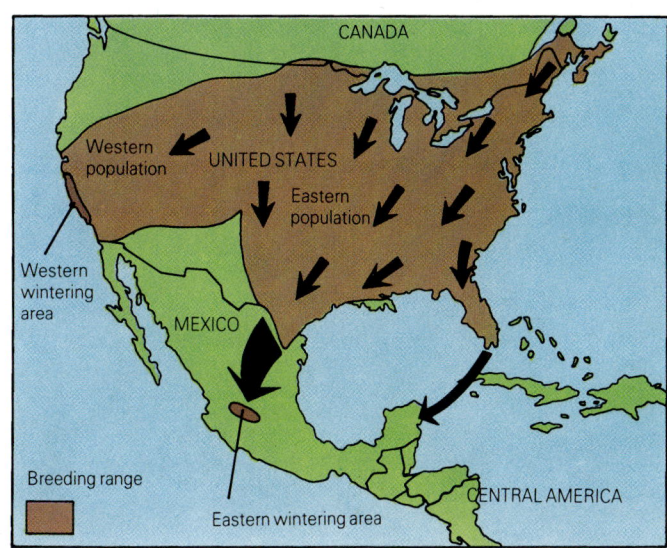

Birds are not the only international migrants. The monarch butterfly migrates from Canada and the United States to spend the winter roosting in the pine forests of the Sierra Madre mountains in Mexico. Since in Mexico there is a great demand for land for growing crops, the people have felled large areas of forest, removing parts of the butterflies' wintering grounds.

The solution to this problem has been to protect some of the butterfly roosts, and to encourage visitors to come to see them. The money which this brings in helps encourage the local people to preserve their natural forests rather than clear them for planting crops.

INTERNATIONAL CONSERVATION

Cooperation across frontiers

Migrant birds are only one example of the need for international cooperation. The state of the world's oceans and rivers, and indeed the whole natural environment, is an international responsibility to be shared by all people.

However, many of the countries where most of the environment is still natural are not able to pay for their own conservation projects. The old industrial countries of Europe, which can afford the money, have very little natural habitat left.

To organize the sharing of the work of conservation, many international bodies have been set up. Some of them are official, such as the International Union for the Conservation of Nature and Natural Resources (IUCN), which is part of the United Nations. Others are unofficial, like Greenpeace, the Worldwide Fund for Nature, the International Fund for Animal Welfare, the World Wildlife Fund, United Animal Nations, and the International Primate Protection League. The International Whaling Commission was set up by all the countries that practiced whaling, in an attempt to prevent whales from becoming extinct.

Above A Hawaiian goose, or nene, at home once more in the mountains of Maui, Hawaii. Birds bred in captivity in England were released into the wild after the numbers of escaped domestic cats had been reduced on the island.

Left The Arabian oryx was another international success story. Animals from the captive herd in the United States were re-introduced to Saudi Arabia, under government protection.

Above Poachers can sell rhino horn for large sums, to be used to make dagger handles, or as an Asian "aphrodisiac."

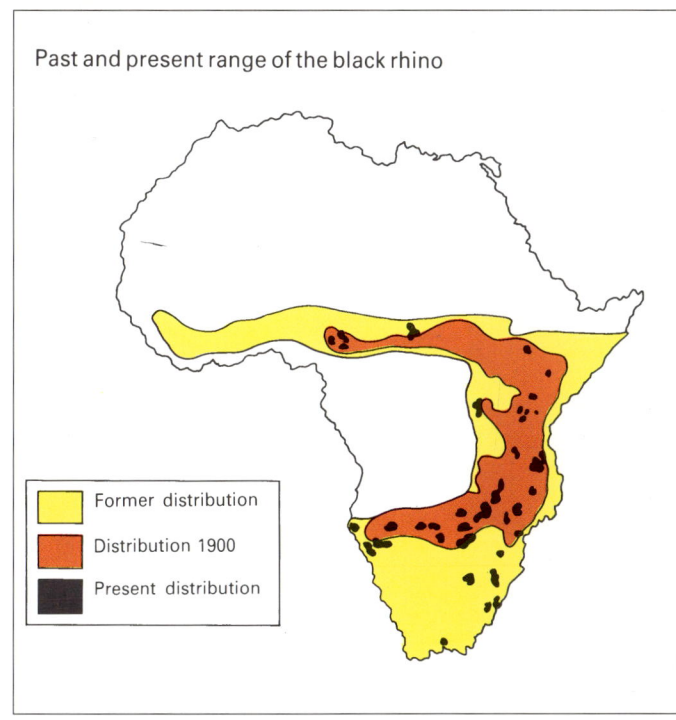

Past and present range of the black rhino

- Former distribution
- Distribution 1900
- Present distribution

Above Since 1980, the black rhinos in Africa have decreased from 15,000 to fewer than 9,000.

Some of these organizations have been very successful. The Fauna and Flora Preservation Society (FFPS) was responsible for rescuing the Arabian oryx from extinction. The animals were bred in captivity in Texas and then re-established in the wild in Saudi Arabia, once the local government had agreed to pass laws to protect them. The Wildfowl Trust in Great Britain saved the Hawaiian goose in the same way.

The Convention on International Trade in Endangered Species (CITES) is an agreement designed to control a trade which threatens the survival of the world's rarest animals and plants. There are still some countries that have not signed it, but more do so every year. The species are considered to be endangered if they are included in the IUCN Red Data Books, which are brought up to date every few years.

INTERNATIONAL CONSERVATION

National and international parks

The first national park in the world was Yellowstone, which covers parts of Idaho, Wyoming, and Montana. Yellowstone National Park was established in 1872. Since then, almost every country in the world has established national parks in the American style.

The first European national parks were set up in Lapland in 1909 and the Swiss Alps in 1914. They were very successful, following a policy of "no harm, no help." This meant that the animals were protected against hunting, but left to look after themselves, even if they were dying of starvation in hard winters.

Many national parks were first set up to save one species of animal. The first of these was in Italy, in 1922, when the king gave his hunting preserve at Gran Paradiso to the state to protect the alpine ibex, which was in danger from overhunting.

Now there is a system of World Heritage Sites. These are places considered to be so valuable that they should not be the responsibility of a single country; instead, they should be run and financed internationally. One example is the atoll of Aldabra, in the Indian Ocean. It is protected for its unique animal and plant life, including the world's last undamaged population of giant tortoises.

The tortoises, like other animals that live on islands, breed more slowly than mainland animals. They have evolved in this way because the island is limited in size. In a national park on the mainland, failure to control the numbers of animals in a park can lead to disaster. In Tsavo National Park, in Kenya, the elephants multiplied until there were too many for the available food and water. They were unable to leave the park to find food elsewhere, and many of them starved. Wise management is needed to keep animal populations at the level which a park can support.

Left Elephants can cause serious damage to trees when food and water run short. In Kenya's Tsavo National Park many ancient baobab trees were destroyed during a severe drought.

Right Yellowstone National Park was dedicated by an Act of Congress on March 1, 1872. Trappers had described geysers and hot springs for years beforehand, but nobody believed them until these sights were confirmed by an official expedition in 1870.

CHAPTER 6: CONSERVATION AND PEOPLE

Waste not, want not

The world's energy needs are so immense that people sometimes wonder when our energy supplies, or resources, will eventually run out. We depend greatly on the fossil fuels (coal, oil, and natural gas), yet we also know that they were formed beneath the earth millions of years ago and cannot be replaced.

To conserve supplies of fossil fuels, scientists are testing alternative sources of energy, which can never be used up and will cause no pollution. These renewable resources include wind power, water power, and (solar) energy from the sun. The challenge to science and technology is to harness such resources efficiently and cheaply. It is also up to governments to put money into research and make decisions for the future.

Minerals are other nonrenewable resources. Some, such as salt, are found in such abundance that we need not worry, but others are in shorter supply. To the factory owner, the word conservation means the reusing, or recycling, of waste materials to make new goods. The recycling industry is growing and will continue to grow in

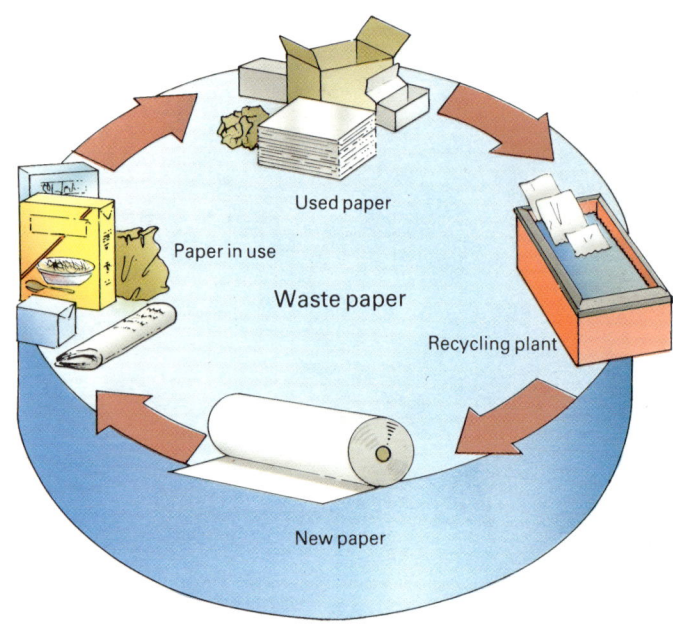

Paper can be recycled once it has been broken down into its original fibers in a chemical bath, and had the ink removed. High-grade paper is the best for recycling.

Left A promising source of renewable energy is the wind, which has been used by mankind for thousands of years to drive ships, and for centuries to turn mills for grinding corn. Modern technology can produce very efficient windmills which generate electricity by turning turbines. This group of windmills is in California, a center of research into alternative energy sources. It can produce enough electricity to power a small town.

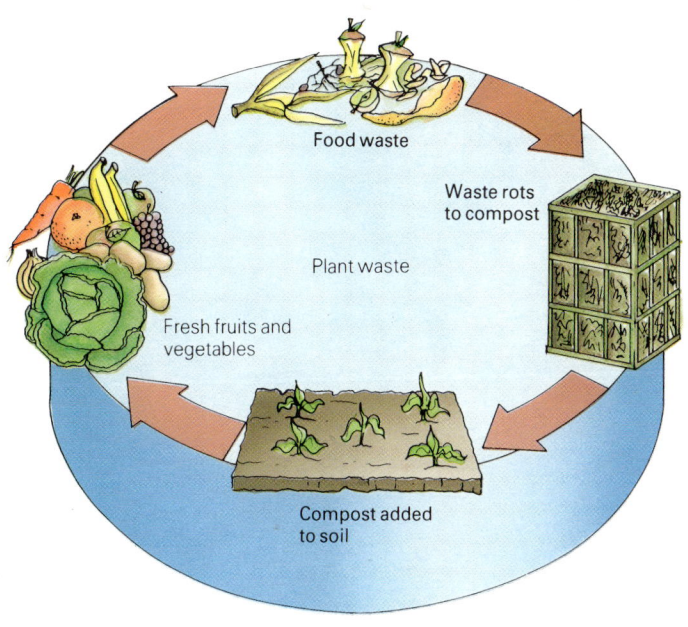

The natural recycling of plant material can be speeded up by composting, in which vegetable waste is collected and allowed to rot until it makes a rich fertilizer.

the future as all types of waste such as metal, glass, and paper are reused again and again.

Many resources—including timber, water, and animal populations—renew themselves naturally. Supplies of renewable resources present no problems as long as the demand for them does not become too great. For example, people now realize that overfishing can lead to serious problems. If too many fish are caught then their population falls quickly. Countries have to agree about how much fish can be caught in each fishing ground, otherwise the fish may die out completely.

Fresh water is another important renewable resource. By the year 2000, at least thirty countries are expected to be short of clean water for irrigation, industry, and household use. The only way to meet the growing demand will be to use water-saving techniques. These will include drip-feed irrigation and the cutting down of water wastage, especially in industry.

Many conservation problems are relatively recent ones. They all have one basic cause—there are too many people in the world.

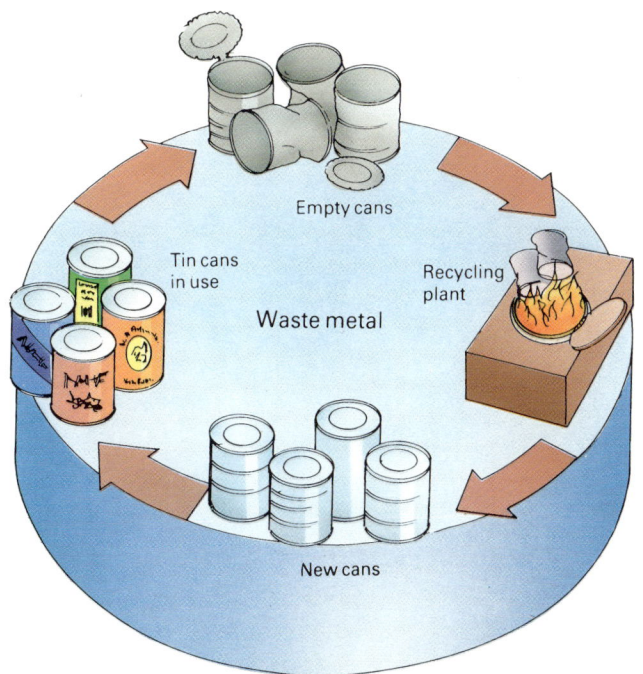

Steel and aluminium cans can be sorted by magnets and melted down to be made into new cans.

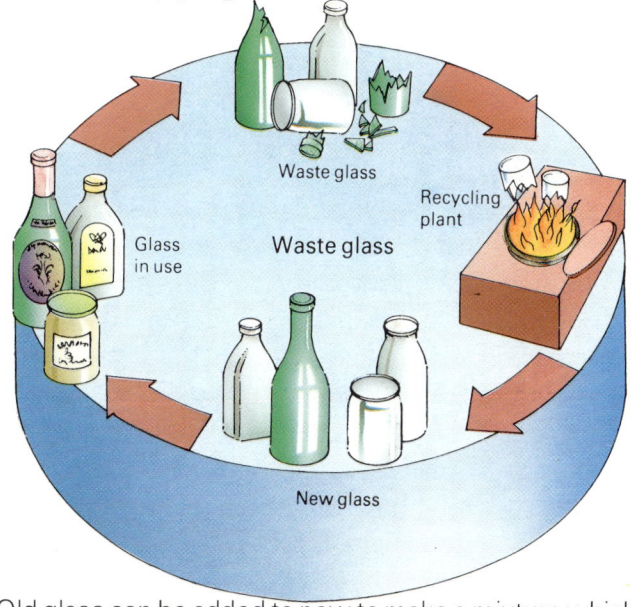

Old glass can be added to new to make a mixture which melts at a lower temperature, saving energy.

CONSERVATION AND PEOPLE

Populations

In the wild, most animal populations live in balance with their natural environment. If their food supply is limited, the population decreases until the balance is restored. Elephants, in India as well as in Africa, are a good example.

Before humans took over the land, elephants wandered far and wide rather like slash and burn farmers. When they had used up the food in one place, they moved on to a new area, returning to the first only when it had had a chance to thrive again. When people began to drive them away from farmland, and later when they were confined in parks, the elephants ran short of food because the vegetation had no time to recover. Many elephants died, until the few that were left could be supported by the available vegetation.

In more than one African national park, it is necessary to slaughter hundreds of elephants every year to maintain a balance between them and their food.

Human populations, too, have outgrown their resources and died of starvation or diseases resulting from overcrowding. With modern communications, everyone in the world is aware of these disasters, and the richer countries hurry to help. Ethiopia and Bangladesh are only two examples of the value of international aid to countries in trouble.

These emergency efforts save the lives of many of the people involved, but they do not solve the long-term problem. In the future, the answer must be to balance the human population with the resources available, not by producing babies that die of malnutrition and other diseases, as at present, but by limiting the number of babies that are born. Unfortunately, there are immense political and religious obstacles to be overcome before this can be achieved.

Overpopulation is the principal cause of all the world's problems of pollution and conservation.

CONSERVATION AND PEOPLE

Antarctica: A clean continent

The problems which now face conservationists arose during the past 200 years. The Industrial Revolution started a wave of pollution which is still spreading across the world. Advances in communications and medicine enabled the human population to multiply and spread as never before. People were not aware of what was happening until in many cases it was too late. Every continent was affected except one. The exception was Antarctica.

The islands in the Antarctic Ocean were discovered first. Sealers arrived at the South Shetlands in 1819, but by 1823 the population of fur seals and elephant seals had been completely destroyed. Whaling was the next mass slaughter, and that is still going on, but on a lesser scale. Krill and fish such as tuna are being exploited today, even though no one knows what damage is being done to the stock.

Antarctica itself was first sighted in 1821, but the first landing was not until 1895. In the early days, the explorers found it hard enough to survive, let alone work there all year round. Now, however, there are many permanent bases dotted around the continent, including one at the South Pole that is like a small town. All the countries that run the bases are members of the Antarctic Treaty.

The Treaty was signed in 1959 by twelve countries, including the United States, Great Britain, Australia, and New Zealand. Since then it has grown to include 35 countries. They agree to hold back from making any claims to the land, while they cooperate in exploration and research. The treaty is due to be reconsidered in 1991.

One of the problems which remains to be settled concerns oil. Although nobody was requested to search for it, geologists have been working there for many years, and some of them believe that there is plenty of oil under the ice-covered rocks.

If the oil reserves were to be exploited, all the problems of pollution, conservation, and conflict that trouble the rest of the world would arrive in Antarctica.

Opposite International cooperation enables scientists to share their knowledge of the Antarctic. Research stations such as New Zealand's Scott Base provide information on geology, meteorology, and biology.

Right Taking samples of the ice by cutting out cylindrical cores provides information about conditions in the past. Deep down the ice is clean, but samples from shallower levels, formed more recently, show signs of the spread of air pollution.

Below Even at the American base at the South Pole, the air is not completely clean.

When a forest has been burned, its recovery can be speeded up by planting the right kind of trees among the ashes.

Glossary

Acid rain Rainwater polluted by waste gases in the air, which upsets the chemical balance of the water and can cause harm to plants and animals.
Algae (singular **alga**) Simple plants that usually live in water. Seaweeds are types of algae.
Atoll A ring-shaped island made of a coral reef.
Bacteria (singular **bacterium**) Very simple, microscopic forms of life.
Biodegradable Describes something that can rot away naturally, by the action of tiny organisms.
Developing World Those countries in which industry has only recently begun to advance.
Ecology The study of life forms in their natural surroundings and the relationships between them.
Environment The surroundings of an animal or plant.
Estuary The broad, tidal part of a river, where it meets the sea.
Eutrophication The over-feeding of streams, ponds, and lakes, due mainly to pollution.
Evolution The gradual change in a species over many generations, as it adapts to changes in its environment.
Extinction The death of an entire species.
Fertile Describes soil that is rich and good for growing plants.
Fertilizer A substance which is added to the soil to make plants grow well.
Finite Something that will eventually come to an end.
Food chain A chain which shows how food energy is passed from one living thing to another, as one is eaten by the next highest in the chain.
Fossil fuel A fuel — such as coal, oil, and natural gas — formed from the fossilized remains of living things.
Ground water Water found in the earth that supplies wells and springs.
Habitat The natural home of an animal or plant.
Irrigation Supplying land with water through pipes, ditches, and canals.
Krill A small sea animal, up to 3 inches long, which looks like a shrimp and is the principal food of whalebone whales, seals, penguins and other sea birds.

Latitude The distance of a place north or south of the equator, measured in degrees; the farther the place is from the equator, the higher is its degree of latitude.
Migrant An animal that travels from one place to another, often over very long distances.
Noise lobby A group of people who try to persuade the government to change the laws controlling noise.
Nutrient Any substance that is nourishing or provides food for a plant or animal.
Overfishing Catching fish faster than they can breed, so that their numbers become less and less.
Pesticide A chemical used to kill insect pests.
Photochemical haze A haze, or mist, formed by certain chemicals that react together in sunlight.
Photosynthesis The process by which plants use sunlight to make food from water and carbon dioxide.
Prairies The vast grasslands of North America.
Radiation The transfer of energy by means of rays of particles or waves; nuclear radiation is energy that comes from the disintegration of atoms.
Radioactive Describes something that gives off nuclear radiation.
Silt Very small particles of soil which settle at the bottom of rivers and lakes.
Smokeless zone An aerea in which fuels that produce smoke cannot be burned, by law.
Soil erosion The wearing away of soil by water or wind, often made worse by the misuse of the land.
Species A group of living things which are alike and able to breed with each other.
Steel foundry A place where steel is made.
Stratosphere A layer in the earth's atmosphere between 7 miles and 30 miles above the surface; the air is very thin and the temperature changes very little.
Temperate zone A region which has a mild climate, between the tropics and polar regions.
Watercourse Any channel of water, including ditches, streams, rivers, and canals.
Water table The level of water in the ground; the depth to which a hole would need to be dug to find water.

Further reading

Benson, Christopher, *Careers in Conservation* (Lerner Pubns., 1974).
Bentley, John and Charlton, Bill, *Finding Out About Conservation* (David and Charles, Inc., 1983).
Brown, Joseph E., *Rescue from Extinction* (Dodd, Mead & Co., 1981).
Cole, Harold, *A Few Thoughts on Trout* (Julian Messner, 1987).
Curtis, Patricia, *All Wild Creatures Welcome: The Story of a Wildlife Rehabilitation Center* (Lodestar Bks., 1985).
Gates, Richard, *Conservation* (Children's Pr., 1982).
Goldin, Augusta, *Water: Too Much, Too Little, Too Polluted* (Harcourt Brace Jovanovich, Inc., 1983).
Kiefer, Irene, *Poisoned Land: The Problem of Hazardous Waste* (Atheneum Children's Books, 1981).
Miller, Christina G. and Berry, Louise A., *Acid Rain: A Sourcebook for Young People* (Julian Messner, 1986, Third edition).
Santrey, Laurence, *Conservation & Pollution* (Troll Assocs., 1985).

Picture acknowledgments

The publishers would like to thank the following for allowing their photographs to be reproduced in this book: British Antarctic Survey 42; Camera Press 21 (top); Central Electricity Generating Board 14 (right); Bruce Coleman Ltd 13 (Hans Reinhard), 25 (bottom/L.C. Marigo), 26 (Jeff Foott), 34 (bottom/Mark Boulton), front cover (inset); Geoscience Features Library 11, 29 (both); The Hutchison Library 8, 22, 23, 25 (top); Mary Evans Picture Library 5 (top); Oxford Scientific Films 18 (G.H. Thompson), 24 (Michael Fogden), 32 (bottom/Maurice Tibbles), 34 (top), 43 (top/Stephen Mills); Planet Earth Pictures 21 (Warren Williams), 28 (Nicholas Tapp), 31 (top and center/Anup Shah), 36 (Ernest Neal), 43 (bottom/Roger Mear), 44, back cover (Richard Matthews); Survival Anglia Ltd 6, 19, 20, 27 (bottom), 30, 32 (top), 33; TOPHAM 5 (bottom), 15, 27 (top); Wayland Picture Library 9 (top) 24; ZEFA 9 (bottom), 16, 31 (bottom), 35, 37, 38, 40, front cover. All illustrations are by the Hayward Art Group.

Index

Acid rain 12
Aerosols 16
Africa 22, 40
Agriculture 6, 26, 28
    chemicals 5, 18
    slash and burn 24
    waste 18
Air pollution 10, 12, 14, 15, 16, 17
Aldabra 36
Alpine ibex 36
Antarctica 17, 42
Antarctic Ocean 20, 42
Antarctic Treaty 42
Arabia 35
Arabian oryx 34, 35
Argentina 28
Artificial fertilizer 18
Atlantic Ocean 20
Australia 16, 22, 24, 27, 28, 42

Bangladesh 22, 40
Berkeley 17
Birds
    flightless 31
    of prey 6

California 17
Canada 12, 32
Canadian maple 12
Cancer 14, 16, 17, 18
    in animals 18
    of the skin 16, 17
Carbon dioxide 12, 17
Carbon monoxide 10
Central America 24
Chemicals 5, 6, 12, 18
Chernobyl 14
Chlorine monoxide 16
Chloroflourocarbons (CFCs) 16, 17
Conservation
    of estuaries 26
    of grasslands 28
    of islands 30, 31
    of rainforests 24, 25
    of wetlands 26
Convention on International Trade in
    Endangered Species (CITES) 35

Darwin, Charles 30
DDT 6
Dieldrin 6, 18, 20
Dodo 31
Dust bowl 22

East Africa 28
Ecuador 30
Elephant 36, 40
Elephant seal 42
Energy resources 38
Erosion 6, 20, 28
    of soil 6
Ethiopia 40
Europe 10, 12, 20, 34
Eutrophication 18
Evolution 30

Fauna and Flora Preservation
    Society (FFPS) 35
Florida 21, 26
Food chain 6
Forest clearance 12, 20, 22, 24, 33
Fossil fuels 12, 14, 15, 38
France 20
Freshwater pollution 18
Fur seal 42

Galapagos Islands 30
Game reserve 31
Germany 10, 20
Giant tortoise 36
Goose 26
    Hawaiian 34, 35

Gran Paradiso 36
Great Britain 10, 12, 14, 20, 34, 35, 42
Greenhouse effect 17
Greenpeace 34

Himalayas 22

Idaho 36
India 22, 40
Indian Ocean 31, 36
Industrial pollution 12
Industrial Revolution 42
Industrial waste 18
Insecticide 18
International aid 40
International Fund for Animal
    Welfare 34
International Primate Protection
    League 34
International Union for the
    Conservation of Nature and
    Natural Resources (IUCN) 34
International Whaling Commission 34
Irrigation 6, 22, 28, 39
Italy 36
IUCN Red Data Books 35

Kakapo 31
Kangaroo 28
Kea 31
Kenya 5, 36
Kiwi 31
Krill 42

Land pollution 6
Land reclamation 26
Lapland 36
Lead 5, 10
    in gasoline 5, 10
    in paint 10
Lead-free gasoline 10

London 10
Los Angeles 10

Manatee 21
Mangrove forest 22
Mediterranean Sea 20
Methane 17
Methyl mercury 18
Mexico 32, 33
Mexico City 10
Migrant 32, 33
Migrating birds 32, 33, 34
Moa 31
Monarch butterfly 33
Montana 36
Mountain forest 22, 33

National park 31, 36, 40
Natural gas 38
New Zealand 30, 31, 42
Nitrate 18
Nitric acid 12
Nitrogen 12
Noise lobby 8
Noise pollution 8
North America 12, 28
North Sea 20
Norway 12
Nuclear
    accident 14, 15
    fuel 15
    power 14, 15
    power station 14
    waste 14, 15

Oil 5, 20, 36, 42
    pollution 20
Otter 6, 18
Overcultivation 22
Overfishing 39
Overgrazing 28

Ozone 16
Ozone layer 16, 17

Pacific Ocean 20
Pampas 28
Pesticide 6, 18, 20
Photochemical haze 10
Plants 25
Plastics 20
Polychlorinated biphenyls (PCBs) 18, 20
Polystyrene 16, 17
Population
    animal 36, 40
    growth of 5, 6, 22, 40
    overcrowding 33, 40
Prairie 22

Radiation 5, 14, 15
Radioactive fuel 14
Radon gas 15
Rainforest 22, 24, 25
Recycling industry 38–39
Renewable resources 38, 39

Sahara Desert 28
Salmon 12, 21
Scandinavia 12
Sea pollution 20, 21
Serengeti plain 28
Sewage 5, 18
Sierra Madre mountains 33
Sloth 25
Smog 10
Smokeless zone 4
Solitaire 31
Songbird 33
South Pole 42
South Shetlands 42
Soviet Union 14, 28
Steppe 28

Stratosphere 16, 17
Sulphur 12
Sulphur dioxide 12
Sulphuric acid 12
Sun (solar) power 38
Swallow 32
Swiss Alps 36

Tetrachloroethylene 18
Texas 35
Thermal pollution 21
Three Mile Island 14
Tidal barrage 26
Tokyo 10
Tributyl tin (TBT) 20
Trichloroethylene 18
Trout 12
Tsavo National Park 36

Ultraviolet (UV) radiation 16, 17
United Animal Nations 34
United Nations 34
United States 10, 12, 14, 17, 21, 22, 23, 27, 33, 34, 35, 36, 38, 42

Wading birds 26
Water pollution 18, 20, 21
Water power 38
Weed killer 6, 18, 29
Wetlands 22, 26
Whaling 42
Wildfowl Trust 35
Wind power 38
World Heritage Sites 36
Worldwide Fund for Nature 34
World Wildlife Fund 34
Wyoming 36

Yellowstone National Park 36, 37